Animal Groups

Mammals

By Dalton Rains

level 1 little blue readers

www.littlebluehousebooks.com

Little Blue House is distributed by North Star Editions:
sales@northstareditions.com | 888-417-0195

Produced for Little Blue House by Red Line Editorial.

Photographs ©: Shutterstock Images, cover, 4, 7, 9, 11, 13, 15, 17, 19, 21, 22, 23, 24 (top left), 24 (top right), 24 (bottom left), 24 (bottom right)

Library of Congress Control Number: 2023902024

ISBN
978-1-64619-810-8 (hardcover)
978-1-64619-839-9 (paperback)
978-1-64619-896-2 (ebook pdf)
978-1-64619-868-9 (hosted ebook)

Printed in the United States of America
Mankato, MN
082023

About the Author

Dalton Rains writes and edits nonfiction children's books. He lives in Minnesota.

Table of Contents

dog

Mammals

A dog is a mammal.

A mouse is a mammal.

mouse

A fox is a mammal.

fox

A whale is a mammal.

whale

A horse is a mammal.

horse

A monkey is a mammal.

monkey

A seal is a mammal.

seal

A bear is a mammal.

bear

A tiger is a mammal.

tiger

Is It a Mammal?

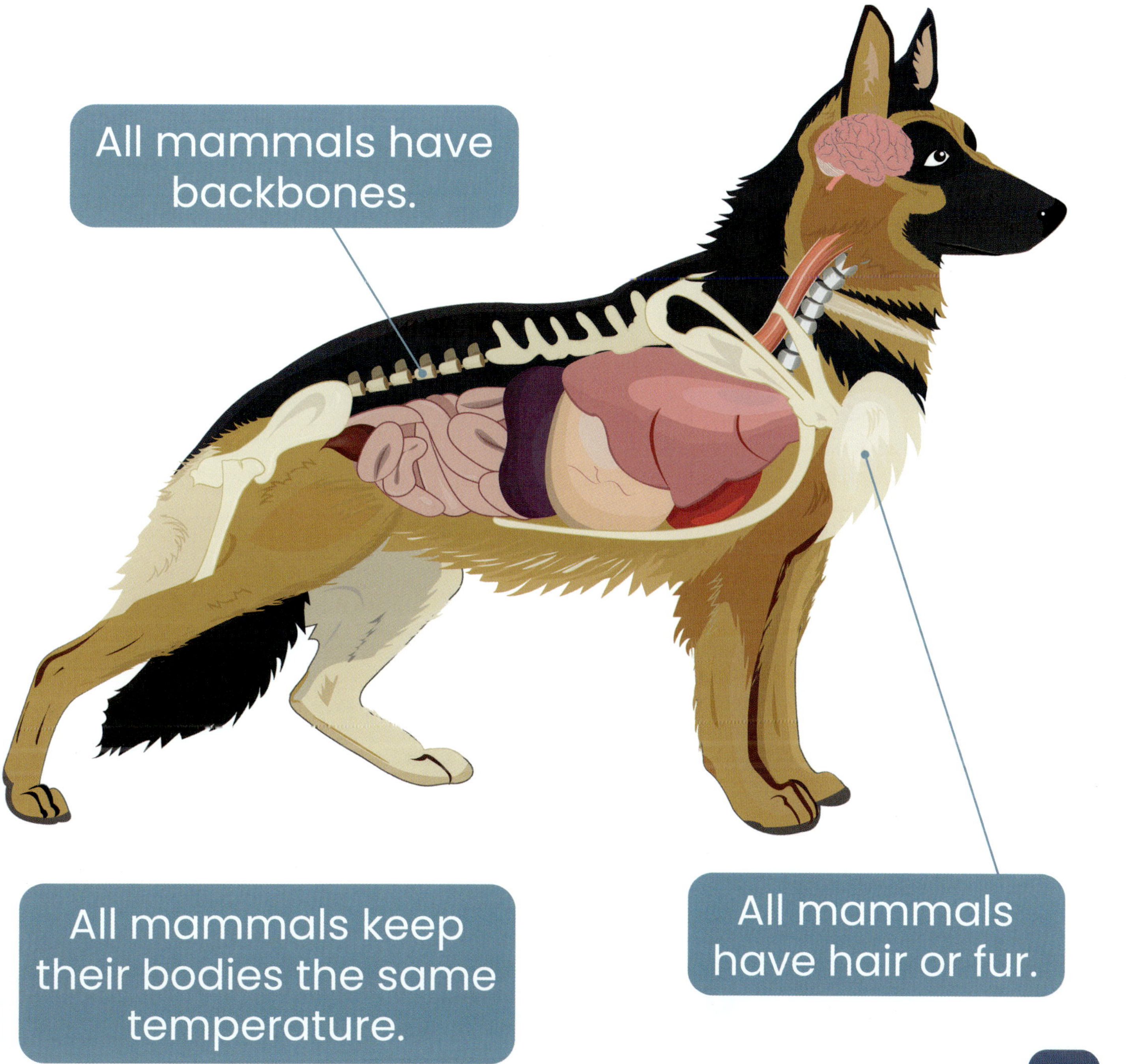
All mammals have backbones.
All mammals keep their bodies the same temperature.
All mammals have hair or fur.

Glossary

bear

mouse

monkey

whale

Index